HANDBOOK OF ENEMY AMMUNITION

PAMPHLET No. 1

GERMAN SHELLS, FUZES AND BOMBS

The Naval & Military Press Ltd

published in association with

ROYAL ARMOURIES

Published by
The Naval & Military Press Ltd
Unit 10 Ridgewood Industrial Park,
Uckfield, East Sussex,
TN22 5QE England
Tel: +44 (0) 1825 749494
Fax: +44 (0) 1825 765701
www.naval-military-press.com

in association with

ROYAL ARMOURIES

In reprinting in facsimile from the original, any imperfections are inevitably reproduced and the quality may fall short of modern type and cartographic standards.

CONTENTS TABLE

FUZES

SEC.		PAGE
1.	Percussion fuze LWMZ.23. Fig. 1	2
2.	Fuze for shells of small calibre AZ.150.Rh.S. Fig. 2	3
3.	Skoda percussion fuze (with and without delay). Figs. 3 and 4	4, 5, 6, 7
4.	Percussion fuze AZ.23.Rh.S. (0·25). Figs. 5 and 6	7, 8, 9, 10
5.	Percussion fuze J.Gr.Z.23.n.A. (0·15). Fig. 7 ...	11, 12
6.	Percussion fuze AZ.23.Rh.S. (0·8) umg. Fig. 8 ...	12, 13
7.	Fuze T. & P. clockwork DOPP.Z S./60. Fig. 9 ...	14, 15
8.	Fuze T. & P. 10 cm. VZ. 21n. Fig. 10	15, 16
9.	Fuze T. & P. 15 cm. VZ. 25	16
10.	20 mm. Aircraft Shell with Fuze AZ. 1502 DWM 11. Fig. 11	17, 18

SHELLS

11	8·8 cm. shell SPRENG-PATR. Fig. 12	18, 19
12.	7·5 cm. separate shell. Fig. 13	20, 21
13.	10·5 cm. streamlined shell with **one** driving band. Fig. 14	21, 22
14.	10·5 cm. streamlined shell with **two** driving bands. Fig. 15	23, 24
15.	15 cm. shell with two driving bands. Fig. 16 ...	24, 25
16.	15 cm. H.E. shell. Fig. 17	26

BOMBS

17.	1 kg. incendiary bomb. Fig. 18	26, 27, 28
18.	1 kg. incendiary bomb (filled). Fig. 19	28, 29
19.	Smoke bomb. Fig. 20	29, 30

1. Percussion fuze LWMZ.23

All dimensions are in mm. unless otherwise stated.

FIG. 1

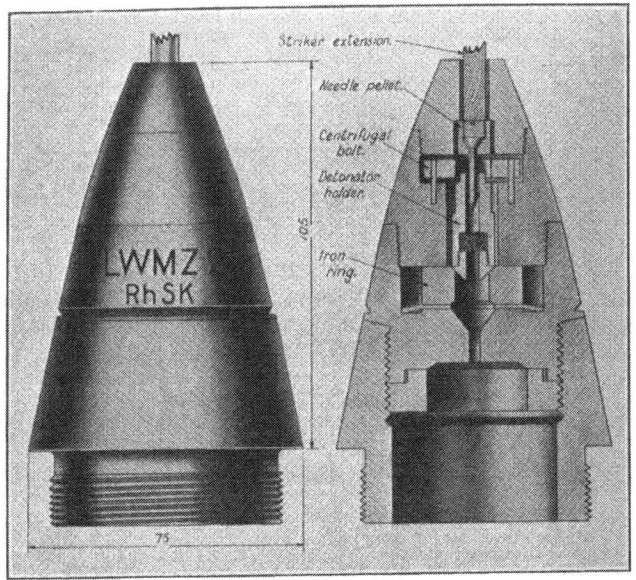

FIG. 1.

This is a direct action fuze in bronze, without delay, actuated by the forcing back, on impact, of a wooden striker extension. Safety is ensured by the arrangement of centrifugally operated safety bolts (*see* description of percussion fuze AZ.23 Rh.S. (0·25)) and the detonator holder is supported on an iron ring as in percussion fuze JGRZ.23.r.A. (0·15).

The marking of the fuze shows that it is used for a light trench mortar bomb and judging by the dimensions of the ogive into which this fuze is screwed the bomb would be of approximately 7·5 cm. calibre.

2. Fuze for shells of small calibre AZ.150 Rh.S.

FIG. 2

This is an extra sensitive fuze with an explosive safety device, armed centrifugally. It is used in shell for 2 cm. aircraft and A.A. guns.

It consists of a fuze body of brass on which is screwed a nose retained by a screw. The body is prepared to take a percussion detonator.

The percussion system consists of :—

(a) A striker kept in a safe position by a centrifugal bolt which fits under a flange below the head of the striker ;

(b) a hammer with an enlarged head to increase the sensitivity of the fuze ;

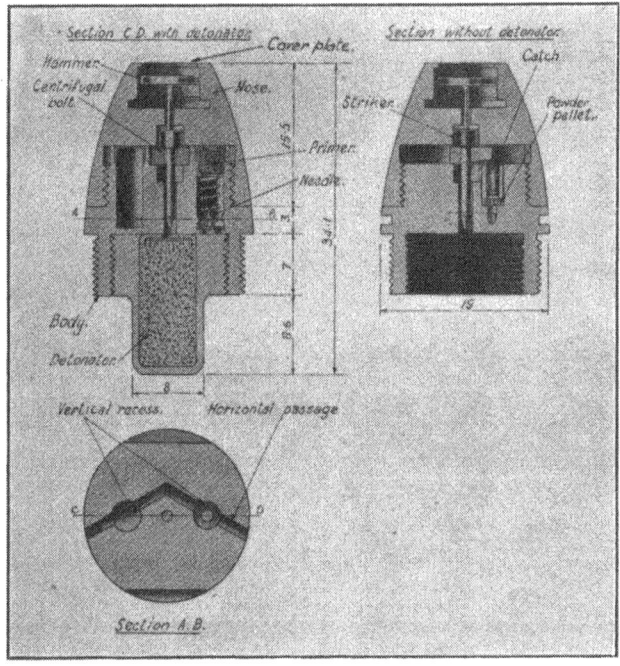

FIG. 2.

(c) a safety arrangement consisting of a catch supported on a pellet of compressed gunpowder. The catch is held with its top rounded bearing surface against the inclined plane of the centrifugal bolt and thus prevents this from moving outwards.

The powder pellet is connected, through two cylindrical horizontal passages, to two vertical recesses, one of which contains a primer supported over a needle by a spring the other identical but empty in the fuze examined.

Action.—On firing the primer sets back on the needle and is ignited. The flash passes through the horizontal passage and ignites the powder pellet. When the pellet is fuzed it frees the catch and thus allows the centrifugal bolt to fly outwards. The striker is then free but creep action due to deceleration in flight and the protection of the cover plate keeps it from the detonator until it is driven in on impact.

3. Skoda percussion fuze (with or without delay)
FIGS. 3 AND 4

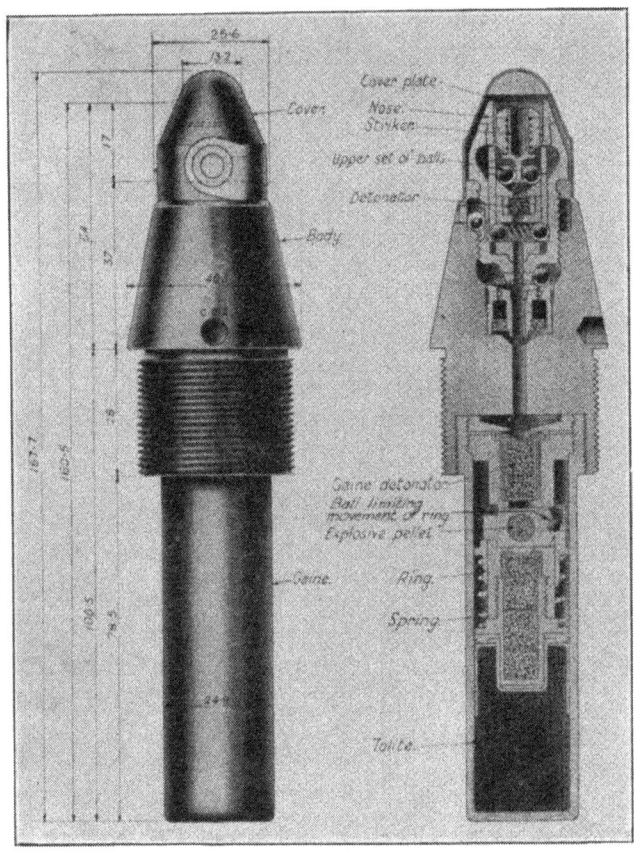

FIG. 3.

This fuze was found in a 100 mm. shell. Fragments of this type of fuze were also found with splinters which appear to be from 75 mm., 83·5 mm. and possibly 150 mm. shells.

The fuze fragments examined carry the Skoda markings as shown in Skoda drawings in the possession of the Schneider Works where the fuze is known as the SKHZR. Skoda drawings show that this fuze can be used for shell varying from 75 mm. to 210 mm. calibre.

Description.

The principal parts of the fuze are :—

(a) Body, (b) percussion mechanism, (c) setting device for delay or instantaneous, (d) the gaine.

(a) The conical steel body is fitted with a brass nose which can be screwed in or out of the body and forms the means of setting the fuze for instantaneous or delay action. The lower part of the body is threaded to take the gaine and the upper part hollowed, screw threaded and sealed with solder. Externally the body is given a protective coating (nature not specified) and fitted with a brass cover, soldered on, which is removed before firing.

(b) The striker is of nickel steel with a head of duralumin screwed on to it. It is protected by a bronze cover plate set into the nose of the fuze and retained with sealing putty. The striker is prevented from striking the detonator before impact by :—

(1) Four polished nickel steel balls which fit between the striker and the detonator holder ;
(2) a creep spring which keeps the striker away from the detonator during flight ;
(3) four castellated notches formed by cutting and bending the top end of a small tube of sheet iron through which the striker passes.

The detonator is secured in the detonator holder by a screwed plug and is contained in a copper tube. It consists of equal parts by weight of fulminate of mercury and inflammable composition. A transverse hole is drilled through the detonator holder to form a seating for the spiral spring which holds the lower set of balls in the circular run in the body.

The detonator holder is seated in a brass tube which has at its lower end two semi-circular notches which retain the lower set of balls in position before firing. The upper part of the tube has three equidistant rectangular notches through which the upper set of balls pass by centrifugal force after firing and two lateral grooves in which the lower set of balls are forced when the brass tube sets backs on firing.

(c) The delay holder contains two parallel delay fillings of amorphous powder and compressed powder. These are retained in position by a brass plate. The holder has a central fire channel which, when the nose of the fuze is unscrewed, is closed by a ball.

To set for instantaneous action (Fig. 3).

The nose is screwed into the body and its lower portion, being coned, displaces the ball from the central channel. The flash from

the detonator can then pass direct through the continuous central channel of the fuze.

To set for delay (Fig. 4).

The nose is unscrewed and the ball, due to gravity, closes the central channel. The flash from the detonator can then only pass through the transverse channels in the base of the nose and so through the delay fillings to the gaine.

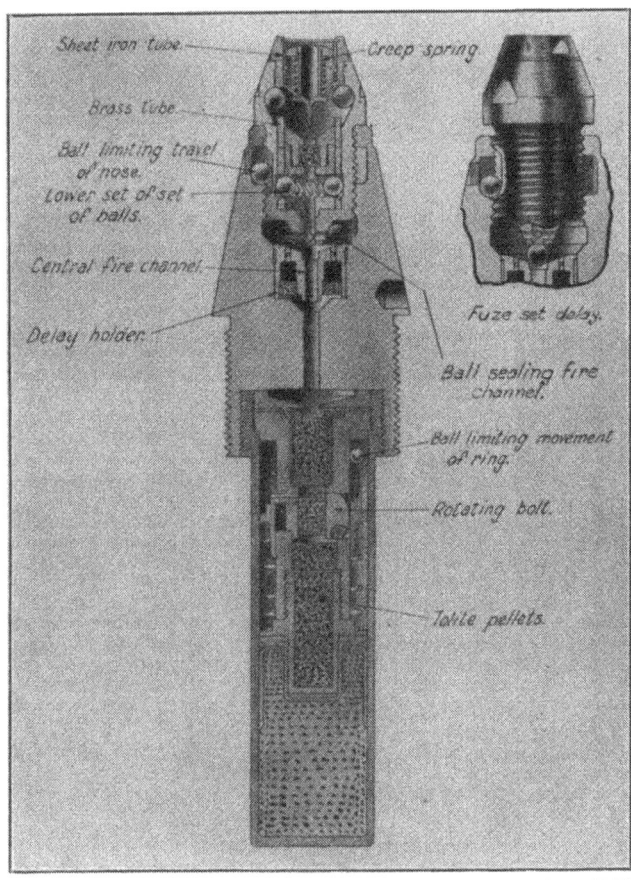

Fig. 4.

After unscrewing and with the ball closing the central channel, the nose of the fuze must then be screwed in again in order to seal hermetically the junction of the ball and channel and to prevent the ball from moving outwards due to centrifugal action in flight.

The amount of unscrewing is limited by a ball which is carried in a seating in the body and is free to move in a groove cut in the threaded portion of the nose.

(d) The gaine, of steel, varnished black, is screwed into the lower part of the fuze. It consists of a detonator, exploder and safety devices.

The detonator consists of fulminate of mercury and compressed Tolite grains carried in a holder and retained by a washer.

The exploders consist of Tolite pellets in varying degrees of compression and are held in a brass tube. The bottom of the gaine is filled with a larger charge of Tolite.

The safety device (Fig. 4) consists of a rotating bolt containing an explosive pellet. The normal position of the bolt is at right angles to the axis of the fuze where it is retained by a ring held in position by a spiral spring. A ball limits the upward movement of the ring. In this position the explosive pellet in the bolt is at right angles to the axis of the fuze, there is, therefore, no communication between the detonator and the exploders.

On firing the ring sets back, compressing its spring. The ball, which is designed to prevent the upward movement of the ring, is now free to move outwards under centrifugal force.

On deceleration, after the shell has left the bore, the ring is forced forwards by its spring and this movement rotates the bolt through 90 degrees. The pellet in the bolt is then in line with the axis of the fuze and completes the transmission of the detonation to the exploders.

Action of fuze.

On firing, the brass tube which retains the lower set of balls in position, sets back. The balls are thrust into the hole of the detonator holder, pass along the grooves and return to the circular runway. Centrifugal force causes the upper set of balls to pass one after the other through the upper notches of the brass tube and come to rest in the runway of the fuze body. Allowing the balls to escape one at a time is an added safety device against prematures.

During flight the striker is kept from the detonator by the creep spring and by the notches cut in the sheet iron tube in the nose. The detonator holder is held by the lower set of balls bearing in the circular runway.

On impact the striker is forced inwards, stripping the notches of the tube and compressing the creep spring. The detonator holder slides through the brass tube overcoming the resistance of the lower set of balls and is carried on to the striker.

4. German percussion fuze, AZ.23.Rh.S. (0·25)

The German percussion fuze AZ.23.Rh.S. with ·25 seconds delay (Fig. 5) is used in the 105 mm. howitzer and probably also in 75 mm. separate ammunition. It is designed to function on impact or graze.

The fuze consists chiefly of a body, needle and needle pellet, centrifugal bolts with spring, detonator pellet with detonator, creep spring, delay mechanism and magazine.

The body, of aluminium, is in two parts, screwed together and secured by a set screw. The upper part has a central channel throughout its length to receive the needle, which is secured in

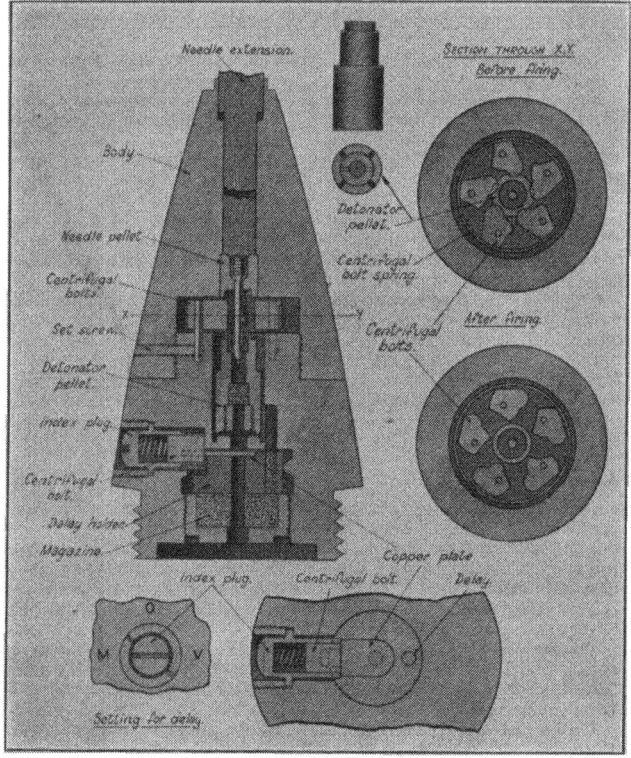

FIG. 5.

the aluminium needle pellet by a securing screw. The needle is fitted with a wooden extension. On the underside of the needle pellet, the central channel is enlarged to house five brass centrifugal bolts, each with its pivot pin. The bolts are kept pressed towards the centre of the fuze by means of a phosphor bronze spring, which

maintains the bolts in such a position that when the fuze is at rest, the needle cannot pierce the detonator.

The lower part of the body contains a brass detonator pellet and detonator, the delay mechanism and a magazine. The detonator is secured in the pellet by a screw having a central fire channel. Four radial slots are cut on the underside of the pellet to ensure that the flash from the detonator reaches the delay channel.

The delay mechanism consists chiefly of a delay holder, index plug with centrifugal bolt and spring, and a copper plate. The holder is pierced by two channels, one central and empty, the other eccentric and carrying the delay. On the upper portion of the holder a recess is cut in which a copper plate can slide. According to its position this plate covers or uncovers the central channel.

The index plug is secured in the body of the fuze by a screwed collar. A cylindrical cavity is formed in the plug to receive the centrifugal bolt with spring and a recess is cut in the plug to receive the copper plate. On the outside of the plug a slot is cut which serves as an index for setting the delay mechanism. If the plug is set in the delay position, the recess does not coincide with the plate, the latter therefore remains in the closed position masking the central fire channel. If the fuze is set to the instantaneous position (Fig. 6) the recess is in line with the plate and the latter is free to move outwards under centrifugal force and so unmask the central fire channel. A brass plate with holes bored to correspond with the delay and central channel is placed on the delay holder and forms an upper bearing surface for the copper plate.

The bottom of the fuze is closed by the magazine, having a central fire channel, which is screwed in and retains in position the delay holder.

Action.

Before firing (Fig. 5).—The needle is separated from the detonator by the centrifugal bolts which are retained in the closed position by their spring. The copper plate of the delay mechanism closes the central fire channel by the pressure from the centrifugal bolt. This position is maintained whether the fuze is set delay or instantaneous. The delay channel is always uncovered.

Thus, even if a failure of the safety arrangements occur and the needle pierces the detonator or the detonator itself fires, the fuze can only function with delay ; the shell, therefore, cannot burst at less than ·25 second's time of flight from the muzzle.

To set the fuze for instantaneous action the slot in the index plug is turned to a position parallel to the axis of the fuze bringing the recess in the plug opposite the copper plate (Fig. 6). For delay action the slot is turned at right angles to the fuze axis opposite the marks M. and V. ; in this position the plug retains the copper plate in the closed position (Fig. 5).

After firing.—The centrifugal bolts swing outwards overcoming the spring thus leaving the needle and detonator pellets free to move towards each other. The creep spring prevents creep action. The centrifugal bolt of the delay mechanism moves outward compressing its spring. If the index plug is in the delay position, the copper plate is held by the plug and thus the central channel remains closed (Fig. 5). If the plug is in the instantaneous position, the

plate is moved by centrifugal force into the slot in the plug and the central channel is thus opened (Fig. 6).

On impact the needle is forced on to the detonator by direct action. On graze the detonator pellet is carried forward on to the needle. The flash from the detonator passes either through the delay channel or the central channel, according to the setting of the fuze, to the magazine and thence to the detonator and exploder in the shell.

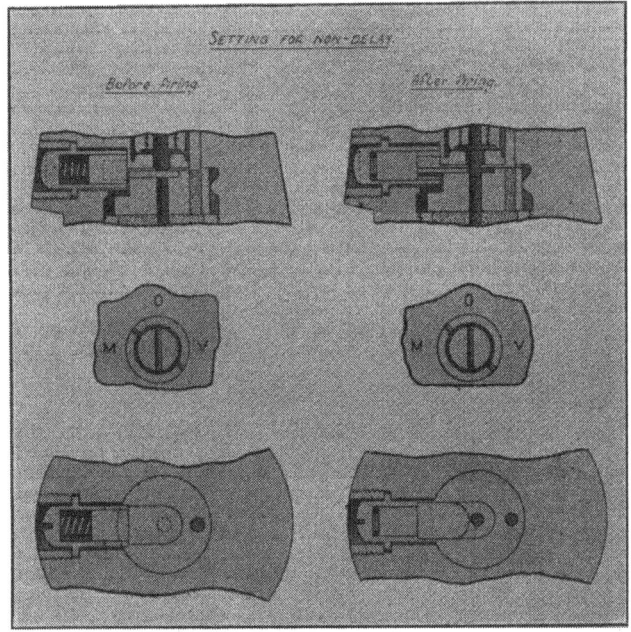

FIG. 6.

Other fuzes operated on the above principle are :—

(a) J.Gr.Z.23.n.A. used with 75 mm. separate ammunition and possibly 105 mm.

(b) A.Z. 23 (0·8) umg. used with 150 mm.

(c) AZ.23.M. (2V.) used with 150 mm.

The only difference between the above types is in the system controlling the delay.

5. Percussion, fuze J.Gr.Z.23n.A. (0·15)

FIG. 7

This aluminium fuze has so far only been found in 7·5 cm. separate ammunition.

In appearance it is similar to Fuze AZ.23 R.h.S. (0·25) previously

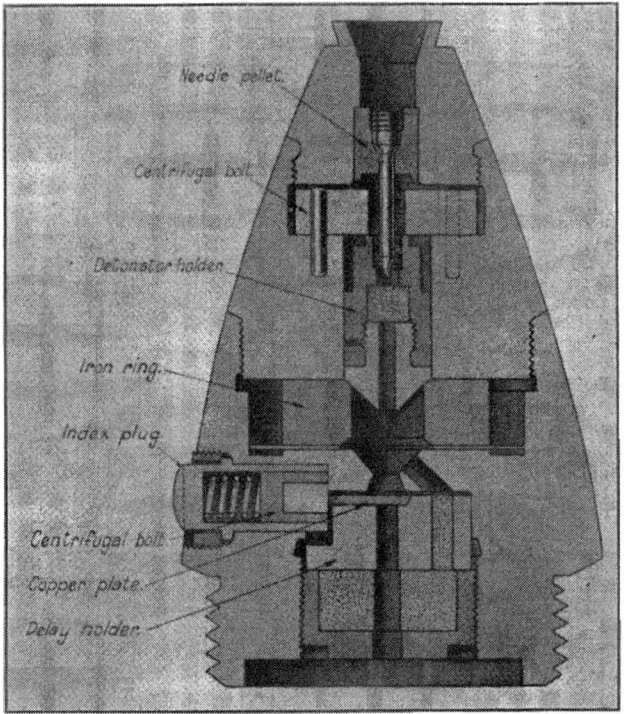

FIG. 7.

described. The only difference is in the dimensions and in the following points of detail :—

(1) The body of the fuze is in three parts instead of two, the three parts being screwed and pegged to each other.

(2) The detonator holder is of a slightly different form with a lower end of truncated cone shape, screwed and pegged into the cylindrical part containing the detonator.

(3) A cylindrical iron ring held by the prongs of a brass washer is held between the centre and lower portions of the body. It supports the detonator holder against the effect of set back and has a central hole communicating with the open central channel and with the channel containing the delay through an oblique passage.

(4) The delay holder is marked 0·15 instead of 0·25.

The delay mechanism and the method of operation are identical with those of the fuze AZ.23 R.h.S. (0·25).

6. Percussion fuze AZ.23 Rh.S. (0·8) umg.

FIG. 8

This fuze has only been found in 15 cm. shells. It is generally similar to the AZ.23 R.h.S. (0·25) previously described—the only difference being in its dimensions and in the delay mechanism.

It consists chiefly of a brass body, a fixed lower ring and an upper setting ring which can be turned by means of notches, using a special key. The lines at O.V. and M.V. can thus be brought to coincide with the lines marked on the body and lower fixed ring.

The fixed lower ring is screwed and pinned to the body. It bears on a shoulder on the setting ring and thus secures it in position. The setting ring on the underside is formed with three distinct bearing planes by means of which the delay mechanism is actuated.

Delay mechanism.

This differs from the AZ.23 (0·25) fuze, in the control mechanism. In this case the copper plate instead of being free in relation to the centrifugal bolt is attached to it by a small pin. Displacement of the centrifugal bolt and consequently the copper plate under the effects of centrifugal force can be prevented by the stem of a detent which is actuated by a spring.

The position of this detent is controlled by the bearing planes of the setting ring. When Part 1 of the bearing planes is over the detent (setting M.V.) as shown in section in Fig. 8, the stem of the latter protrudes into the space in which the centrifugal bolt is positioned and prevents its movement. The central channel is thus closed and the fuze can only function on delay.

When Part 2 or 3 of the setting planes (setting O.V.) is brought over the detent, the latter is free to move slightly longitudinally and, under pressure from its spring, its tip is withdrawn from behind the centrifugal bolt. The bolt can then move outwards and compress its spring thus withdrawing the copper plate and leaving the central channel of the fuze clear.

It will be observed that the setting cap has three bearing planes, although only two are required in this fuze. The reason for this is that the setting cap is identical with that used with the AZ. 23 umg. M.2.V. (which is similar to the AZ.23 (0·8) umg.) in which the three settings (no delay, 0·2 delay, 0·8 delay) are used. In this case, the three planes are necessary. It is, therefore, probably with a view to securing standardisation that the fuze described here has three planes as in the corresponding setting cap of the AZ.23 umg M.2.V.

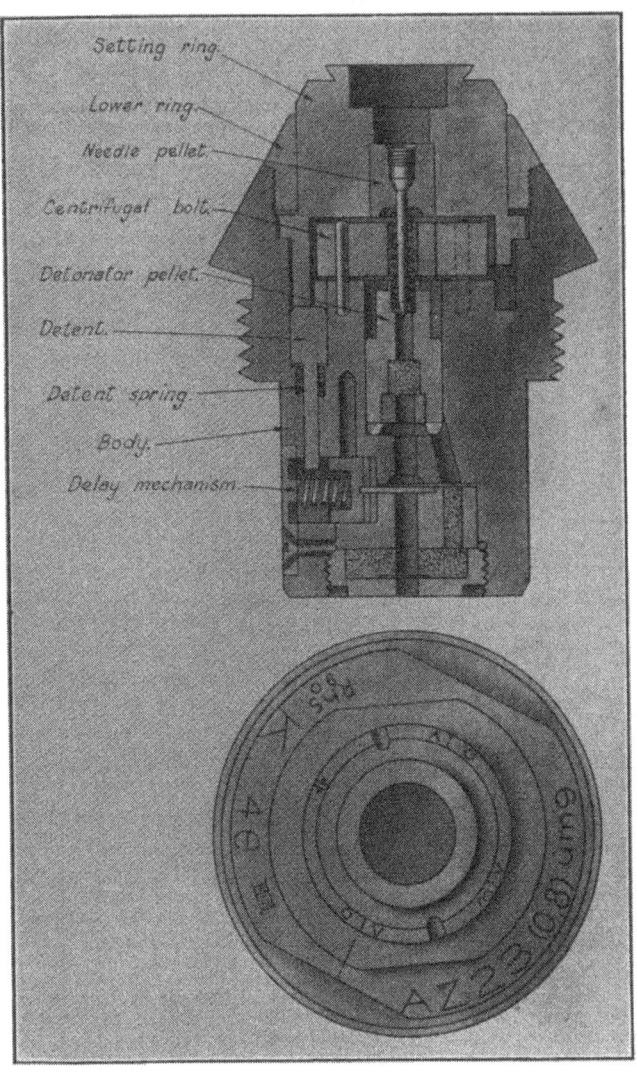

Fig. 8.

7. Fuze T. & P. Clockwork DOPP. ZS./60s.

FIG. 9

This fuze is used for air burst ranging. The body is made of duralumin with the clockwork movement in bronze. A number of these fuzes have been recovered, but they have been very deformed

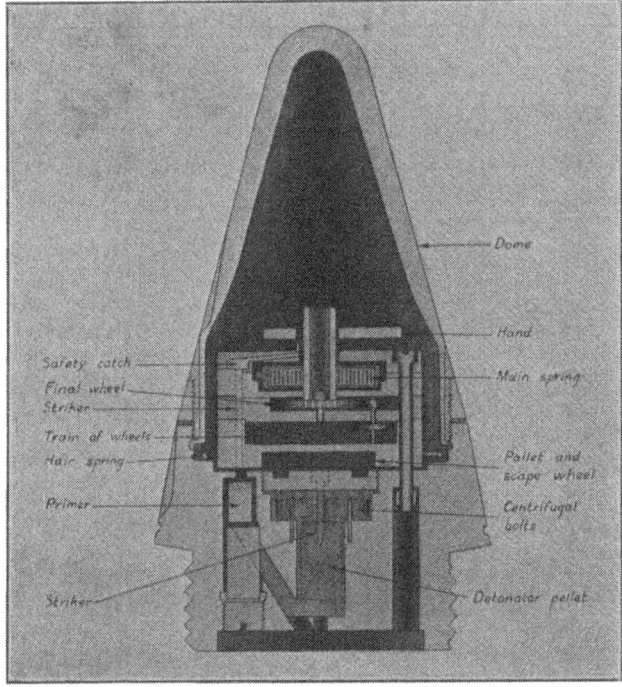

FIG. 9.

so that it has not been possible to give full details of the mechanism. This seems to be approximately as follows :—

Time mechanism.

The clockwork movement is attached to the body of the fuze by three screws, arranged at intervals of 120°. It appears to resemble clockwork fuzes used by the Germans in 1914-1918.

It consists of :—

(a) a pallet and scape wheel. The timing of the oscillation is controlled by a hair spring of 0·6 mm. spring steel

15 cm. long, which passes round the outside of the clockwork movement.

(b) the scape wheel driven by a train of wheels with a ten-tooth pinion and 56-tooth wheel engaging a final wheel (40 teeth) fixed to the main spring drum.

(c) the main spring, ratchet wheel (60 teeth), striker, safety catch, hand, etc.

By using the slots in the body and in the head of the fuze, the dome is turned until the fuze is set at the correct fuze length.

After the primer has been struck the flash is carried to a detonator which is probably the same as that in the percussion system.

Percussion mechanism.

The percussion system is below the clockwork movement. It was very broken in every case and consequently is only shown in the Fig. by dotted lines.

It consists of a safety mechanism of four centrifugal bolts of ordinary type and a detonator pellet which, on graze, is carried forward on to the striker which is fixed.

Some fuzes instead of being engraved DOPP. ZS./60s. had the same marking with the " s " barred out. Factory markings, such as Rh. S. 1936 have been found.

8. Fuze T. & P. 10 cm. VZ.21n.

Fig. 10

This fuze which is constructed partly in bronze and partly in steel, is of the combustion type with two time rings filled with powder burning composition, the top ring being fixed and the bottom movable. The bottom ring revolves between two felt washers and is graduated from 0 to 241 sub-divided into tenths the fuze being set by revolving the bottom ring until the required graduation is opposite an index line on the base of the fuze. To protect the powder filling against moisture, a cap is placed over the fuze and is soldered to the base. Presumably some form of waterproofing the time rings is also used.

The time mechanism consists of a primer case supported by a safety clip and held in position by a safety pin which is withdrawn at the moment of loading.

A percussion pin is fixed in the body of the fuze below the primer.

Percussion mechanism.

Only fragments of the percussion system were found. These consisted of a striker fitted with an aluminium head. No manufacturing marks were visible.

Action.

On firing the primer case sets back crushing the safety clip and strikes the percussion pin. The flash is carried to the upper time ring which burns for a period dependent on the setting, then to the lower graduated ring and eventually by the diagonal channel to the magazine in the ordinary way.

The gas from the burning composition in the time rings escapes through large vents, one in each time ring.

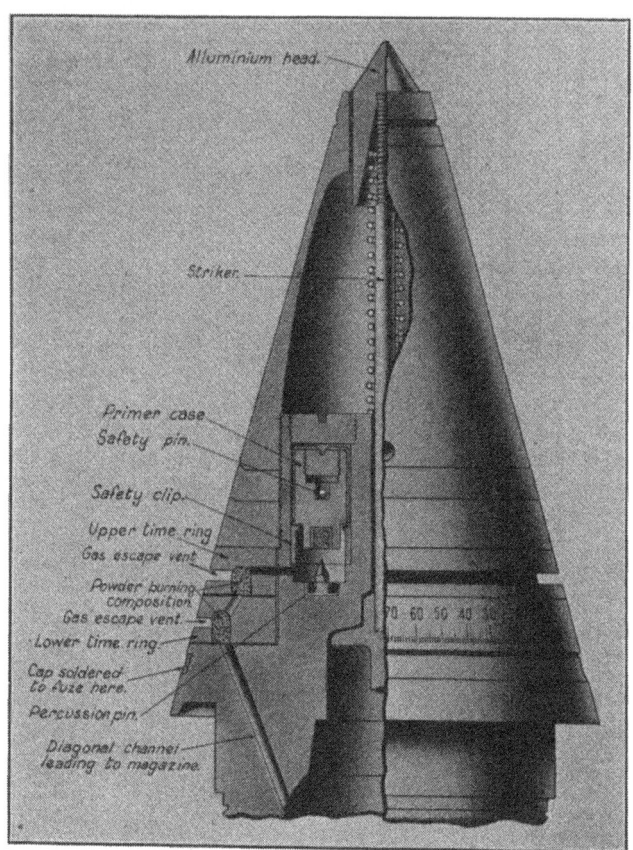

Fig. 10.

9. Fuze T. & P. 15 cm. VZ.25

This fuze is generally similar to the 10 cm. VZ.21n., differing only in the following particulars :—

(a) the diameters of the screw threads and of the base plates are slightly larger and the overall length is slightly less.
(b) the graduations on the bottom ring are from 8 to 245.
(c) a few details, particularly the shape of the striker, are very slightly different.

10. 20 mm. aircraft shell

Fig. 11

This H.E. Shell has been found on various occasions after aerial combat between Messerschmitt and French Aircraft.
It consists of a steel body in three parts, comprising :—

(a) A cylindrical body.

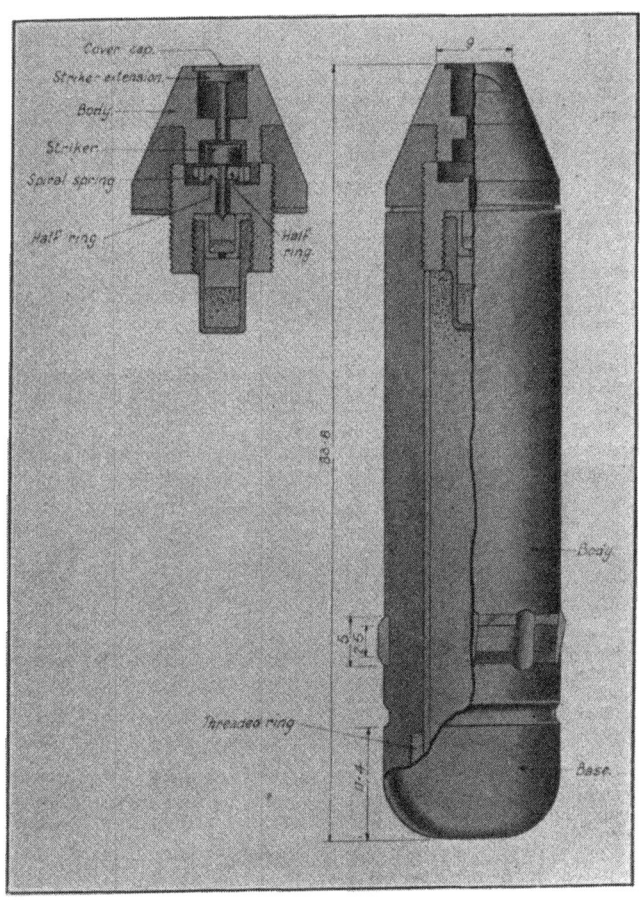

Fig. 11.

(b) A practically hemispherical part which forms the base of the shell.

(c) A threaded ring screwed inside (a) and (b) and acting as a connection between them.

At the upper part of the base a cannelure is formed which is probably for indenting the cartridge case.

The body of the shell is painted yellow as far as the cannelure, the part fitting inside the case being unpainted. It is filled with a pink-coloured explosive.

The shells recovered were fitted with either a fuze marked AZ.1502 DWM.11, or a fuze marked EKZ.dr.C./30.

Fuze EKZ.dr.C./30 is an instantaneous nose fitted on a tracer shell and carries inside it a brass tube which contains the tracer composition. This is mainly lead and manganese and is in the form of a white powder.

Fuze AZ.1502 DWM.11

Fig. 11

This D.A. fuze screws into a ring carried in the head of the shell and is fixed with three punch marks.

The body is of steel coated with brass and the nose of the fuze is closed with a cover cap.

The arming system consists of a spiral spring rolled around two half rings, which retain the striker in position.

On firing the spiral spring unrolls under the effect of centrifugal force thus releasing the two half rings, which in turn free the striker. Both spring and half collars remain inside the fuze.

During flight " creep " action keeps the striker clear of the detonator. On impact the striker extension is driven in actuating the striker.

11. German 8·8 cm. shell spreng—Patr.

L/4·5 (KZ)

All dimensions are in mm.

Fig. 12

This H.E. Shell has a screwed-in base with parallel walls and two driving bands of either copper or bimetallic fitted into undercut grooves near the base. Two cannelures are formed below the driving bands for the purpose of securing the cartridge case to the shell.

For identification purposes the groove immediately below the driving band is painted black and the grooves between and above the bands are painted yellow.

The shell is filled T.N.T. wrapped in a varnished cardboard container insulated from the inside walls by four sheets of white paper and from the base by two superimposed washers of black cardboard.

This shell is used in the 8·8 c.m. FLAK 18 A.A. gun and is fitted with one of the following fuzes :—

(a) against aerial objectives : mechanical clockwork fuze UZS./30 ;

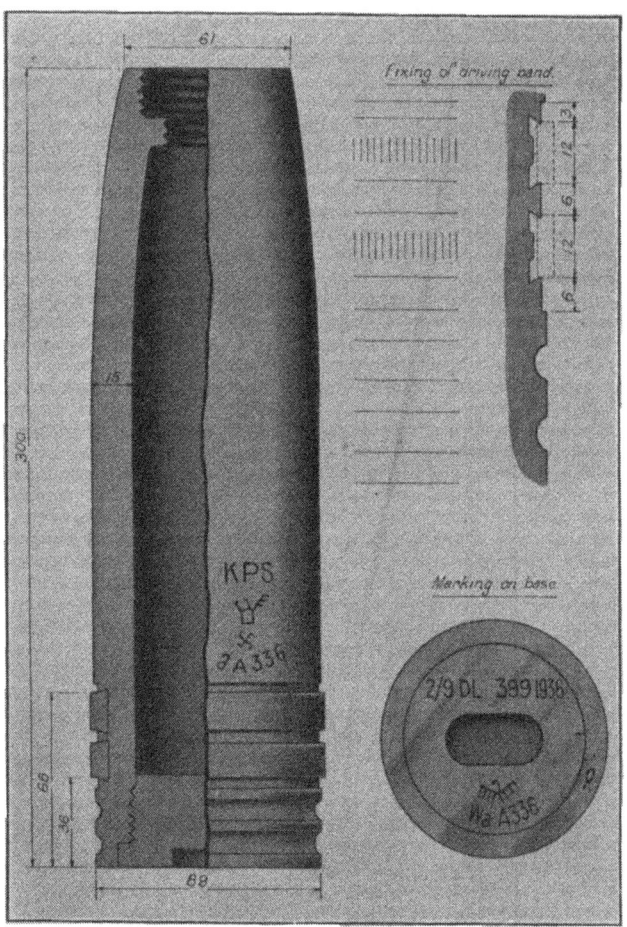

Fig. 12.

(b) against land objectives (armoured vehicles) percussion fuze AZ.23 (0·15) or time fuze Zt.Z.S.30 (0·5,000).

The weight of shell is 9 kgs., cartridge 14·7 kgs. and the charge 2·3 kgs.

12. 7·5 cm. separate shell
FIG. 13

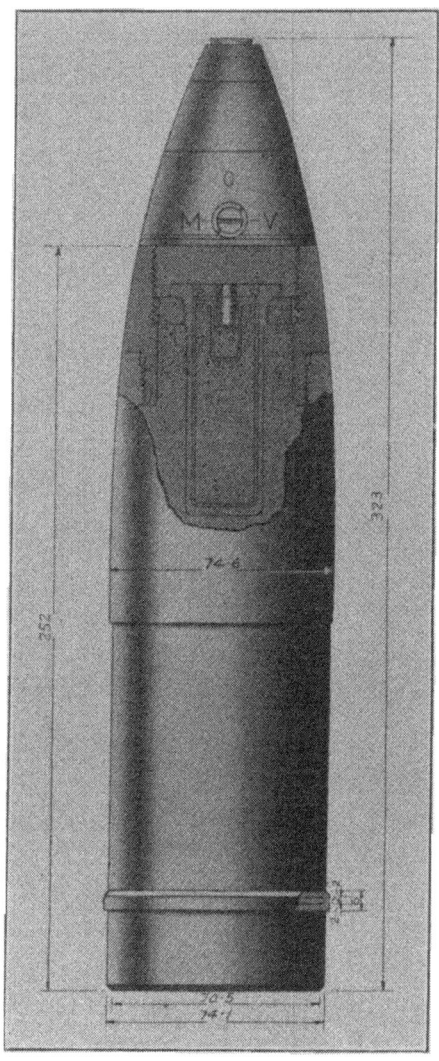

FIG. 13.

This shell, of steel, has a screwed-on head and a parallel base fitted with one driving band.

The H.E. filling, probably T.N.T., is brought to the level of the ring securing the exploder container. The inside walls are protected by varnish.

The shell recovered was painted Green and fitted with fuze J.Gr.Z.23 n.A. (0·15) R.h.S. 1937 previously described.

The explodering arrangements are identical with those used in the 10·5 cm. shell (Fig. 14).

It is very probable that the above shell is that used by the Light Infantry gun, which fires, as also does the heavy cm. infantry gun, a shell with a separate propelling charge. On the other hand the firm of Rheinmetall has made a 7·5 cm. L./25 gun with split trail, and a 7·5 cm. L./42 gun (long range) mounted on a split trail carriage of the 10·5 cm. model 18 light howitzer. The above shell could have been fired by one of these weapons or by a 7·5 cm. FK. 16n.A.

Weight of shell without fuze 5·2 kg.

13. Streamlined 10·5 cm. shell with one driving band

Fig. 14

Fig. 14 shows the principal features of a German 10·5 cm. H.E. shell fitted with one driving band and a percussion fuze AZ.23 with ·25 in. delay.

The shell is streamlined and fitted with a screwed-on head. The driving band consists of an inside portion of mild steel or iron and an external portion of copper. The method used to prevent the band slipping round the body of the shell is by simple knurling. The adhesion of the copper to the iron is so good that twisting of the band on firing does not cause any separation between the two metals. The reason is not clear whether the two-metal driving band is due to shortage of copper or as a remedy for the tearing off of driving bands.

The shell filling is probably T.N.T. In some cases the filling is effected by pouring, the shell being protected internally by varnish, and in other cases by block charges, each contained in a carton cover, a plastic substance being used to make a joint between the covers and internal surface of the shell. The shell is filled up to the level of the ring securing the exploder container.

The steel exploder container is secured by means of a steel securing ring, the inner thread of which is screwed to the container, and the outer thread to the fuze hole of the shell. These two components are treated externally with a protective coat of black preservative the nature of which is not specified.

An exploder of **PICRIC ACID** contained in a cylindrical casting of tinned brass fits into the container.

The detonator containing the cap is housed in the upper part of this casting. The detonator consists of a small copper thimble pierced by five holes (nature of filling not stated). The thimble is closed by a washer, the central hole of which is closed by a paper disc. The detonator is held in position by a leather washer crimped into the edge of the cylindrical brass casting.

The same type of exploder appears to be used in 75 and 150 mm. H.E. shells.

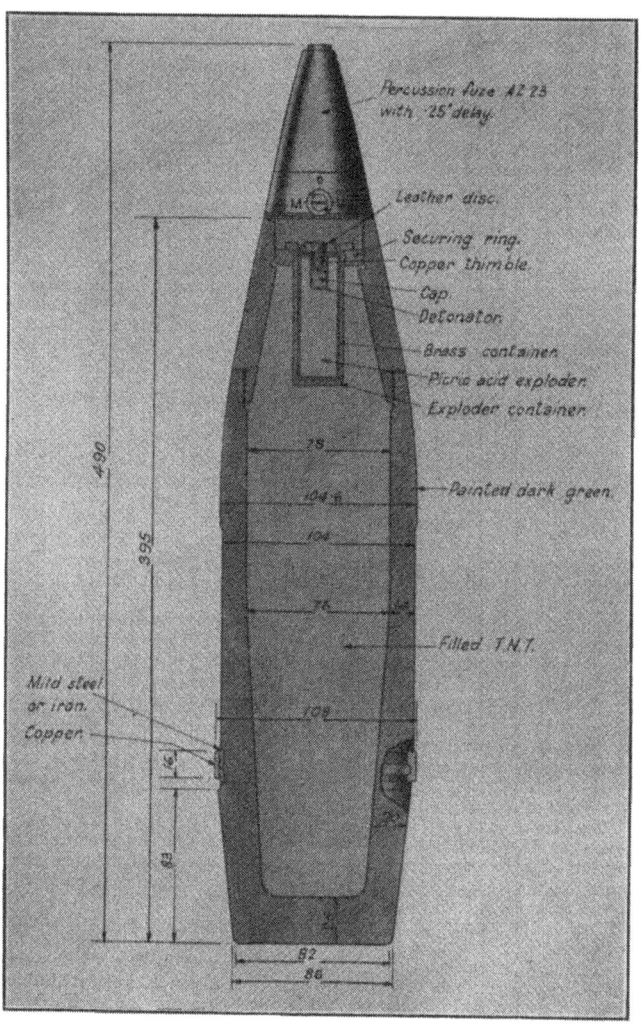

Fig. 14.

The driving bands of certain projectiles of 105 and 150 mm. are in some cases of copper only and in this case the method used to prevent the band slipping is in the form of a checker pattern.

14. Streamlined 10·5 cm. shell with two driving bands
FIG. 15

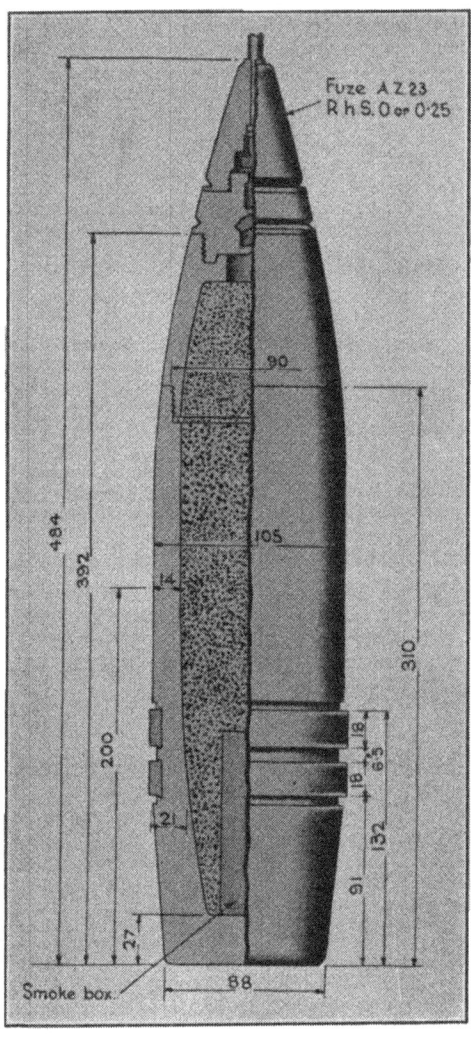

FIG. 15.

This shell is fitted with two driving bands 18 mm. wide, a screwed-on head and has a solid base.

It is filled with a yellow explosive contained in a cardboard wrapper which is separated from the metal by a layer of plastic compound. For ranging shell the base of the filling contains a smoke box of red phosphorus.

Shell of this type which have been recovered were fitted with aluminium fuzes AZ.23 R.h.S. 1938, with delays 0 and 0·25 or without indication of delay.

15. 15 cm. Shell with two driving bands

FIG. 16

This shell has a solid head and a screwed-in base, the latter having two spanner notches. The base is streamlined.

The explodering arrangements consist of a gaine in the form of two superimposed cylinders. The upper and larger end screws directly into the fuze hole of the shell and inside it fits the cylindrical lower end of the fuze. The exploder container fits into the lower cylinder. The exploder is identical with that used in the 10·5 cm. shells.

The shell is varnished inside and the filling, probably T.N.T., comes up to the level of the lower edge of the gaine. In certain samples recovered the filling is done in loose cartridges with cardboard wrapping.

The total weight of the shell without the fuze is 42·7 kg.

The fuzes which appear to be used with this shell are :—

   (i) either the steel fuze AZ.23 umg M.2 V. R.h.S./90, operating without delay or with 0·2 or 0·8 delay ; or

   (ii) the brass fuze AZ.23 (0·8) umg R.h.S.K. 5°, operating without delay or with 0·8 delay.

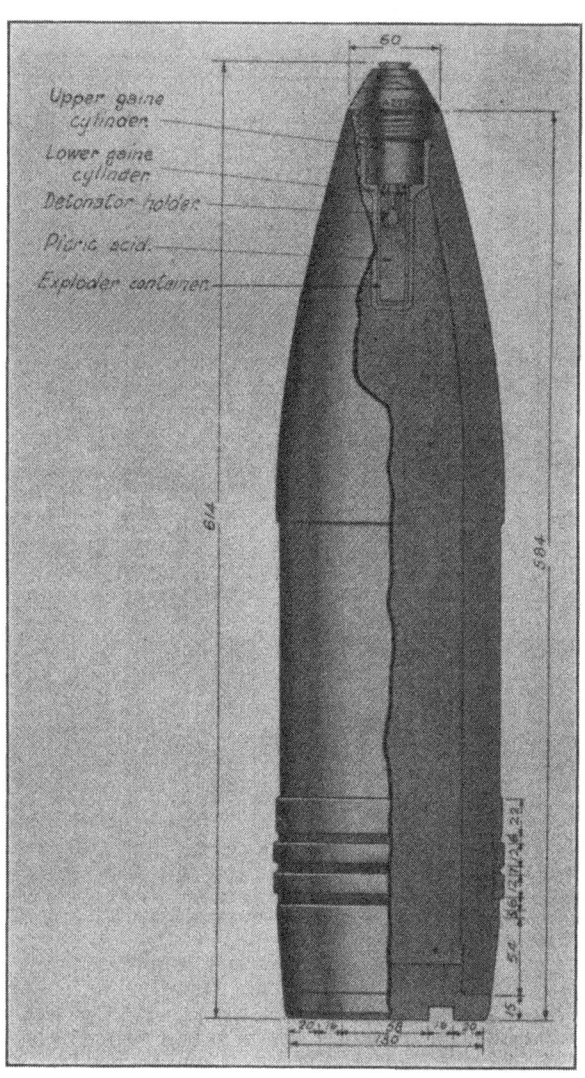

Fig. 16.

16. 15 cm. H.E. Shell
FIG. 17

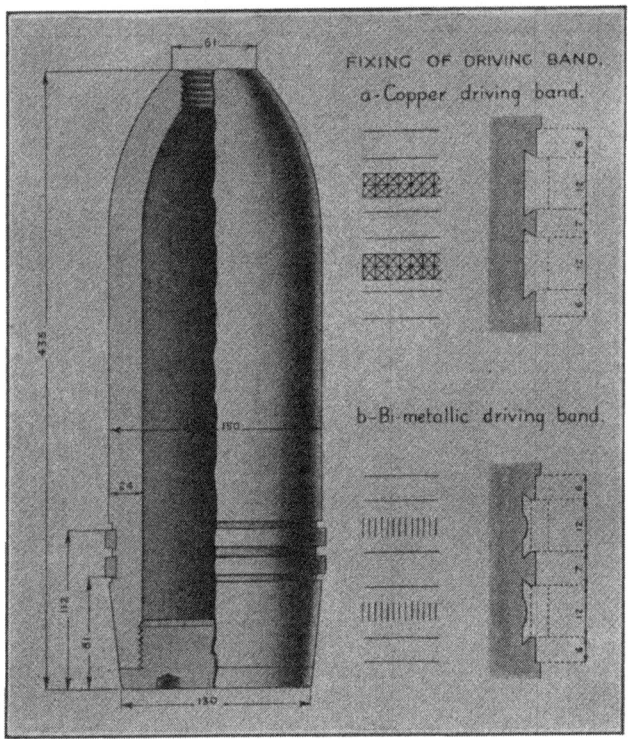

FIG. 17.

This shell is streamlined, with parallel walls, a screwed-in base piece and a head of about 1½ calibres which is threaded to take a nose fuze. The base is fitted with two driving bands of either copper or bimetellic.

17. 1 kg. incendiary bomb
FIG. 18

This type consists of a thick walled tube 9 inches long and 2 inches in diameter, made of an alloy of magnesium with a small proportion of aluminium. One end of the tube is fitted with a tail 5 inches long. The tube is filled with a priming composition of the thermit type. The bomb is fitted with an igniter which may be either in the nose or tail end of the tube.

The bomb weights about 2 lbs. 2 ozs. and, with the exception of a few ounces in the tail and igniter, there is no dead weight, the whole being incendiary material. The bomb functions on impact, a needle in the igniter being driven into a small percussion cap which ignites the priming composition. The bomb does not explode.

It should be noted that, although this bomb is often called a thermit bomb or a thermit electron bomb, the main incendiary

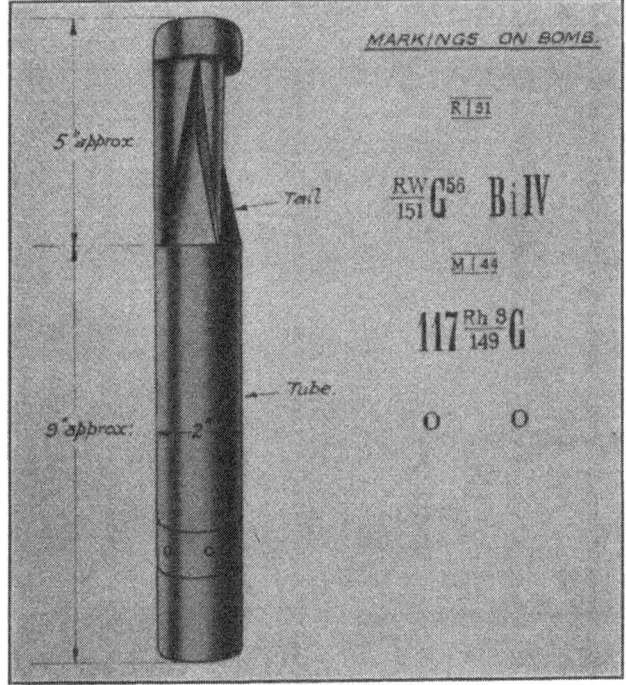

Fig. 18.

agent is not the thermit composition but the magnesium tube, which is not in itself readily inflammable. The priming composition burns for 40-50 seconds at a temperature of about 2,500° C., and its great heat serves to melt and ignite the magnesium tube. The molten magnesium burns for 10 to 15 minutes at a temperature of about 1,300° C. It may remain active for as long as 20 minutes and will set fire to anything inflammable within a few feet.

During the first 50 seconds or so, while the priming composition is still burning, the bomb looks very violent. Jets of flame are emitted from the vent holes and pieces of molten magnesium may be thrown as far even as 50 feet. After the first minute the bomb becomes less active because the magnesium tube melts and the pressure within is released.

The thermit composition contains its own oxygen and so cannot be extinguished by smothering, but the magnesium must get its oxygen from the air or surrounding materials in order to burn.

18. 1 kg. incendiary bomb (filled)
FIG. 19

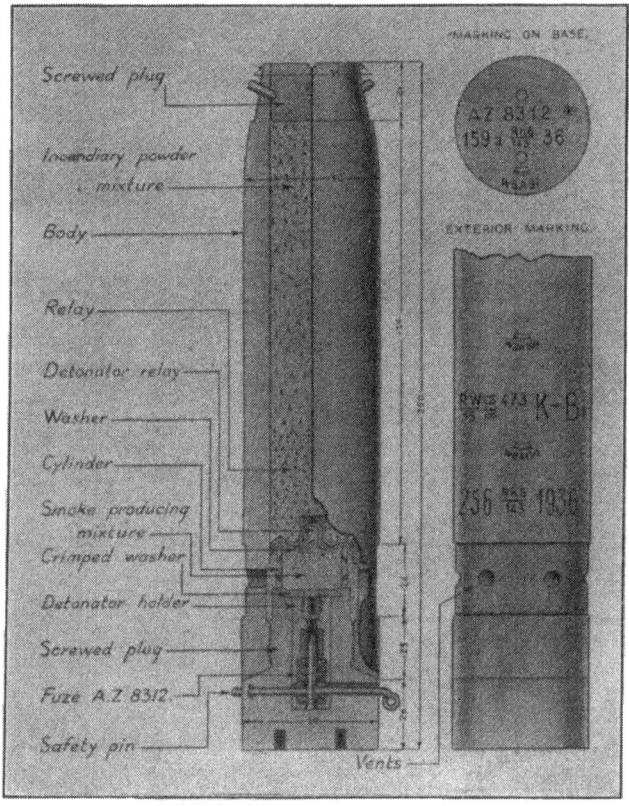

FIG. 19.

This bomb consists of a cylindrical body made of an alloy of magnesium and aluminium with traces of silicon and is closed at its upper end by a screwed plug. On the shoulder of the bomb, three pegs, which also serve to lock the plug, hold an extension which has not been identified but is probably the tail. The base is closed by a plug which houses the percussion fuze AZ. 8312.

This fuze consists of a striker, an arming spring and a detonator holder. The whole is sealed and held in position by a crimped washer. A safety pin keeps the needle clear of the detonator. The igniting arrangements consists of a detonator relay kept away from the plug by a washer and a cylinder with vents drilled in it. The whole is covered by a mixture which appears to be smoke producing. The vents in the body of the bomb serve for the release of the gases. These vents are normally closed by a strip of Chattertons compound.

The main filling is in two parts. The first, of small quantity, in contact with the detonator, serves as a relay. Its composition is not known. The main filling is an incendiary powder mixture of iron, aluminium, etc.

19. Smoke Bomb
Fig. 20.

This bomb takes the form of a cylinder resembling in shape and size a tin of preserves.

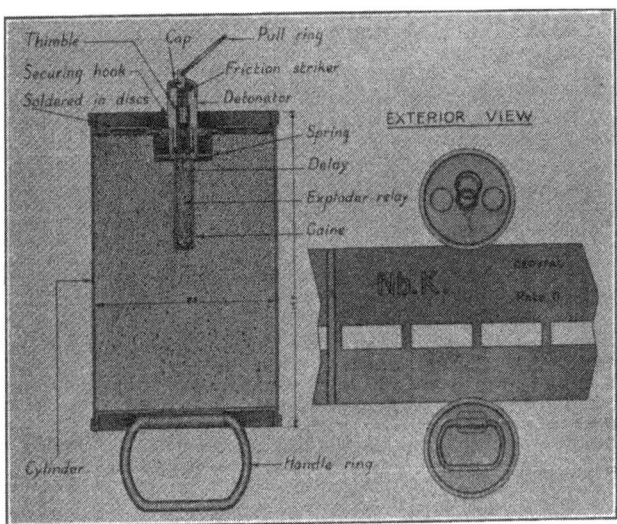

Fig. 20.

Into one end of the cylinder is screwed a detonator adapter of brass fitted with a pull ring with securing hook. At the other end is fitted a handle ring, placed eccentrically so that, when released, it falls flat on the bottom of the tin.

The cylinder is of tinned plate, pressed and soldered on the sides and ends and contains the smoke producing block. A spring limits the movement of the block in the cylinder and also prevents it coming into contact with the gaine. Two holes at the top of the cylinder are closed by discs soldered in.

The exploder is screwed into the brass adapter which also houses the gaine, soldered in.

The firing mechanism consists of :—

(a) A pull ring held in a safe position by a securing hook.
(b) A friction striker which is attached to the pull ring and passes through a small thimble of red copper which contains the detonating mixture. The striker consists of a piece of fine wire coiled in a spiral at each end of the thimble so as to allow for the straightening out of the wire when it is pulled. The wire is kept central at the time of pulling by the cap which can only move axially on the adapter and which a recess prevents from rotating.
(c) A tinned brass tube closed at one end and plugged at the other by a small brass piece containing the delay. This tube also contains the exploder relay.

Action.— When the pull ring is free from the securing hook, pulling out of this ring stretches out the striker and fires the detonator. The flash is transmitted by the delay to the exploder relay, then to the smoke mixture and the smoke pours out by the holes at the top, the closing discs of which have then become unsoldered.

Marking.—The tin is painted dark green with a white band around the centre having four equal gaps.

Note.—The Germans had in 1917 three types of smoke producing weapons :—

The model N.T. (Nebel-Trommel) weighing 115 kilos.
The model N.L. (Nebel-Topf) weighing 69 kilos in three loads of 23 kilos.
The model N.K. (Nebel-Kasten) which could be carried in two loads of 17 kilos. each.

These weapons were not actuated as in the bomb described above by hand pulling of a striker, but by turning through 180 degrees by means of a handle the drum containing the materials to be mixed. The model N.T. put out for 20 minutes a thick white fog which might last up to half an hour.

According to the regulation instructions on the method of use, this fog was in no way toxic but simply caused irritation in the throat. On the other hand the liquid contained in the apparatus was very caustic and would have caused injury to the eyes. Consequently, the man who was operating the weapon had to wear special glasses.